Wildlands Fire Management:
Federal Policies and Their Implications
for Local Fire Departments

Reported by: Christina Rossomando

This is Report 045 of the Major Fires Investigation Project conducted by TriData Corporation under contract EMW-90-C-3338 to the United States Fire Administration, Federal Emergency Management Agency.

Homeland
Security

Department of Homeland Security
United States Fire Administration
National Fire Data Center

U.S. Fire Administration Fire Investigations Program

The U.S. Fire Administration develops reports on selected major fires throughout the country. The fires usually involve multiple deaths or a large loss of property. But the primary criterion for deciding to do a report is whether it will result in significant "lessons learned." In some cases these lessons bring to light new knowledge about fire--the effect of building construction or contents, human behavior in fire, etc. In other cases, the lessons are not new but are serious enough to highlight once again, with yet another fire tragedy report. In some cases, special reports are developed to discuss events, drills, or new technologies which are of interest to the fire service.

The reports are sent to fire magazines and are distributed at National and Regional fire meetings. The International Association of Fire Chiefs assists the USFA in disseminating the findings throughout the fire service. On a continuing basis the reports are available on request from the USFA; announcements of their availability are published widely in fire journals and newsletters.

This body of work provides detailed information on the nature of the fire problem for policymakers who must decide on allocations of resources between fire and other pressing problems, and within the fire service to improve codes and code enforcement, training, public fire education, building technology, and other related areas.

The Fire Administration, which has no regulatory authority, sends an experienced fire investigator into a community after a major incident only after having conferred with the local fire authorities to insure that the assistance and presence of the USFA would be supportive and would in no way interfere with any review of the incident they are themselves conducting. The intent is not to arrive during the event or even immediately after, but rather after the dust settles, so that a complete and objective review of all the important aspects of the incident can be made. Local authorities review the USFA's report while it is in draft. The USFA investigator or team is available to local authorities should they wish to request technical assistance for their own investigation.

For additional copies of this report write to the U.S. Fire Administration, 16825 South Seton Avenue, Emmitsburg, Maryland 21727. The report is available on the Administration's Web site at http://www.usfa.dhs.gov/

U.S. Fire Administration

Mission Statement

As an entity of the Department of Homeland Security, the mission of the USFA is to reduce life and economic losses due to fire and related emergencies, through leadership, advocacy, coordination, and support. We serve the Nation independently, in coordination with other Federal agencies, and in partnership with fire protection and emergency service communities. With a commitment to excellence, we provide public education, training, technology, and data initiatives.

TABLE OF CONTENTS

EXECUTIVE SUMMARY

The United States Fire Administration (USFA) has prepared this report for fire service professionals interested in keeping abreast of developments affecting forests and wildland firefighting policies of the Federal government and the fire problem in the rapidly growing wildland/urban interface areas. This paper looks at two of the major wildfires in 1988 and the policy issues that surrounded them: the Greater Yellowstone Area fires and the "49er" fire that caused more than 20 million dollars in damage in the Gold Rush country of Nevada County, California.

Through the years, U.S. Forest Service (USFS) and National Park Service (NPS) fire management policies have taken many forms. An early belief in the importance of suppressing all fires in wildlands evolved into current policies that reflect a basic belief that fire is natural to many wildlands and plays a vital role in the ecology of those lands. The Greater Yellowstone Area (GYA) fires provided a crucial test under extreme conditions of NPS and USFS wildfire management policies in effect in 1988.

The Interagency Fire Management Policy Review Team was formed to review fire management policies and their implementation during the 1988 fire season. The team found that fire management policies were basically sound, but many of the plans devised for individual forests, parks, and wilderness areas did not conform to national policy guidelines. They recommended that steps be taken to strengthen and ensure operational compliance with those policies. The panel also found serious shortcomings in many of the fire models, or "prescriptions," specifying in what areas and under what conditions natural or ignited fires will be allowed to burn in order to meet land management objectives.

In addition to strengthening implementation of fire management policy, the team recommended:

- Fire management plans for national parks and forests be improved by:

    (a) developing joint agency fire management plans, agreements or addenda to existing plans for those areas where fire could cross administrative boundaries;

    (b) including additional factors in the comprehensive set of criteria to be used in deciding whether or not to allow natural ignitions to burn as prescribed fires;[1]

    (c) clearly describing the decision process and factors to be addressed before a fire is declared a prescribed natural fire;

---

[1] Existing criteria already included: consideration of land management objectives for the area, historic fire behavior, natural role of fire, proposed degree of suppression, expected fire behavior, acceptable suppression techniques, adequate buffer zones, smoke management and effects on adjacent landowners. New criteria added include consideration of the effects of cumulative drought conditions on fire behavior, availability of firefighting resources and considerations of the regional and national wildfire situation, to name a few.

A "natural ignition" is defined as a naturally occurring fire, such as a fire started by a lightning strike. "Prescribed fires' may be ignited by managers or naturally occurring fires that are allowed to burn under specific "prescribed" conditions to achieve established management objectives. Thus a "prescribed natural fire" is a naturally ignited fire that is allowed to burn because it is burning under specific, acceptable conditions and meets park or forest management objectives.

(d) including criteria to be used in declaring a prescribed fire a wildfire;

(e) clearly identifying areas that need protection from fire, such as developments within or adjacent to wilderness and park boundaries, and including within fire management plans actions that are to be taken to protect them, i.e., hazard fuel reduction or installing fuel breaks;

(f) clearly stating the management objectives being addressed by the prescribed natural fire program; and,

(g) clearly describing the process to be used to ensure adequate public involvement and coordination with local governments.

- Regional and national contingency plans be developed to constrain prescribed fires under extreme conditions.

- Agencies consider planned ignitions to complement prescribed natural fire programs and to reduce hazard fuels.

- More consideration be given to the effects of fire policies in surrounding communities and steps be taken to encourage more public participation in the development of individual fire management plans.

- Additional research and analysis on weather, fire behavior, fire history, and fire information integration be carried out so that future fire management programs can be more effective in reducing the risk of wildfires.

The panel also cited wildland/urban interface issues as an area in need of further study. Specifically, it recommended further study to determine the current and future effects of residential and commercial development on the ability to design and implement prescribed fire programs, and to examine the interrelationship between fire management plans and local planning and zoning functions.

ACTIONS SINCE 1988

Since then, both the NPS and the USFS have taken numerous steps to implement the recommendations of the Fire Management Policy Review Team. Among the actions taken are the following:

- A detailed review was immediately undertaken of all fire management plans in national parks and forests that included use of "prescribed natural fire" (naturally-ignited fires allowed to burn under certain "prescribed" conditions).

- Both the NPS and the USFS imposed a temporary freeze on their "let burn" policies until these fire management plans were revised to meet current policy and new requirements.

- The NPS instituted a daily certification process that requires verifying the availability of adequate firefighting resources to assure that prescribed natural fires will remain within prescriptions. If not, they are to be declared wildfires and suppression actions taken immediately. NPS also reviewed its policies regarding the use of prescribed burning (e.g., planned ignitions) in place of prescribed natural fire.

- The National Wildfire Coordinating Group (NWCG) was ordered to develop criteria for a national contingency plan for controlling natural fires under extreme conditions, and regional offices of the NPS were ordered to develop regional contingency plans.

- The USFS revised its fire management manual (*Forest Service Manual 5100 – Fire Management*) to implement the recommendations of its 1989 in-house "Task Force Report on Prescribed Fire Management Criteria." This report enumerated detailed recommendations for the Forest Service's policy and guidance documents spelling out how policies are to be implemented.

- The NPS issued a revised fire management policy manual (NPS-18) in early 1991. This completes a revision process that began before the 1988 wildfire season and was expanded as a result of the fires.

WILDLAND/URBAN INTERFACE

The wildland/urban interface is the area where relatively untouched wildlands and residential areas meet. The influx of new residents into these interface areas is likely to increase the incidence and severity of the fires and to have major implications for firefighting efforts.

Fire service organizations ought to be aware of the steps they can take to prevent and fight wildfires in wildland/urban interface areas. These include:

- Overall management plans for parks and national forests are revised periodically. Fire service organizations in areas near national forests, parks, and wilderness areas can affect wildfire management policies and practices by taking advantage of opportunities for public participation in the development of fire management plans by testifying at public hearings or submitting comments on proposed plans. Hearings and comment periods are announced in *The Federal Register* and advertised in local newspapers, and notices are posted in U.S. Post Offices. Nearby USFS and NPS offices are listed in local telephone directories under "U.S. Government."

- Fire departments should be aware of measures to reduce fire dangers to homes and other structures in wildland/urban interface areas, such as using fire-resistant roof materials and requiring defensible space free of vegetation in areas adjacent to structures.

- Fire departments in wildland/urban interface areas often need to take aggressive steps to implement and strictly enforce local week abatement and brush clearance ordinances.

- Fire departments in wildland/urban interface areas can use prescribed burning as a valuable, cost-effective fire management tool to clear heavy brush and accumulated fuels.

- Firefighters in volunteer fire departments and fire protection districts in interface areas need to become more knowledgeable about wildland fire prevention and firefighting methods, such as compressed air foam systems.

THE FIRES

The 1988 wildfire season is one that will not soon be forgotten. When the smoke had cleared, 72,084 wildfires had burned 4,309,316 acres in the western States and Alaska, making 1988 the worst wildfire season since the 1920s, according to the Boise (Idaho) Interagency Fire Center. Firefighting costs were estimated at over 500 million dollars, and seven lives were lost.

Millions of Americans watched daily as television showed flames ripping through much of the Nation's favorite national park. As the Greater Yellowstone Area fires threatened a number of nearby communities, two million acres burned out of control in Alaska and other devastating wildfires caused damage in populated areas near wilderness, such as the Gold Rush country of California, where more than 100 homes were lost.

The Clover-Mist and North Fork Fires in the Greater Yellowstone Area

The Greater Yellowstone Area fires of 1988 involved the largest fire suppression effort ever undertaken in the United States and focused attention on Federal fire management policies in wilderness areas. The USFS estimates that more than 25,000 firefighters fought the fires over more than three months. At its height, an estimated 9,500 firefighters and 117 aircraft were involved in the suppression effort.

The Greater Yellowstone Area encompasses 11.7 million acres of land in Montana, Idaho, and Wyoming and includes parts of two national parks and six national forests, as well as State lands, Federal reservations, unreserved public domain and other lands. During 1988, there were 249 fires in this vast area, 201 of which were suppressed at less than 10 acres and 13 of which became major blazes that changed the face of the Yellowstone area for generations to come.

Together, the Greater Yellowstone Area fires burned an estimated 1.41 million-acre area in and around the park and destroyed 67 structures, including 19 cabins in the park and four houses just outside of it. Original estimates that the 13 major Greater Yellowstone Area fires burned approximately 988,000 acres within the park itself were revised downward in January 1990 after park officials had time to carefully analyze satellite photos. The final estimate is that about 793,800 acres burned about 36 percent of the total area of Yellowstone National Park. Still, the magnitude of the loss is staggering, especially when you consider that in the previous 100 years, only 200,000 acres of Yellowstone had burned.

The following is a synopsis of the two largest of the Greater Yellowstone Area fires: the Clover-Mist fire that burned in the northeastern part of Yellowstone National Park and was contained on October 7; and the North Fork fire that threatened the Old Faithful village at various times during late summer and early fall of 1988.

The Clover-Mist Fire

The fire that became known as the Clover-Mist began as seven separate lightning-ignited natural fires on July 11. By July 21, four of the fires burned together and were declared wildfires. Suppression action was taken on the southern and eastern flanks. At this point, the fires covered 31,500 acres.

Strong winds July 24-26 swelled the fire to 46,800 acres and by July 28 the fire had grown to 68,035 acres. Between August 13-21, three more fires joined with the main blaze. On August 20, the fire made an eleven-mile run driven by tremendous winds, and 46,500 acres were added. It soon involved 170,700 acres.

Between September 8-9, winds of up to 50-75 miles per hour (mph) developed and fanned fires burning in Sunlight, North Crandall, Jones Creek, Papoose, and Squaw Valley. Flames consumed structures in Crandall, just outside of the park. Damage included three residences, six outbuildings, 13 trailers, one store, and three vehicles.

Snow in higher elevations on September 9-10 provided a break that allowed firefighters to gain ground, and by September 13 the tide had turned. The Clover-Mist fire was finally declared contained on October 7, after it had grown to approximately 319,000 acres in size.

The North Fork Fire

The North Fork Fire began on July 22, reportedly by a logger who carelessly discarded a cigarette in the Targhee National Forest north of Yellowstone National Park. It was immediately classified a wildfire and was fought from the start, but the blaze escaped the initial attack and grew quickly. On July 23, 25-40 mph winds spread the fire from the forest land into Yellowstone National Park. It was now 340 acres in size, and a firefighting team was immediately called in to suppress it.

The fire took a major run on July 23 and headed toward the Old Faithful area. Protection action was taken for structures in Old Faithful Village while the flames were about 12 miles away. That meant that a containment strategy was used on the western and northwestern flanks so that the main suppression efforts could be concentrated at the Village, where fuel reduction actions were taken and fire lines were set up. By July 26, the fire moved to within six miles of the Village. Cooling temperatures allowed firefighters two days of direct suppression efforts.

On July 30, strong winds came into play and the fire spread to 11,300 acres. Major runs on August 11 and 12 caused it to grow to 34,120 acres. Strong winds on August 15 swept the blaze across Madison Junction, but no buildings were lost. Norris was evacuated on August 17, and structure protection actions were taken on August 18 as the fire grew in size to 69,000 acres. By the next day, it involved 72,410 acres.

August 20 was a day that firefighters in Yellowstone are not likely to forget. Forty mph winds with stronger gusts caused a major fire run toward the northeast. The North Fork fire, now forty miles long and 104,000 acres in size, split into two heads on August 25, one north and one south of Norris Junction.

The northern branch of the fire was 77,800 acres in size. Pushed by northerly winds on the 27th, it posed a renewed threat to Old Faithful Village and parts of the Targhee National Forest. U.S. Army forces were put on the fire lines. The North Fork fire moved downslope on September 1 and threatened the town of West Yellowstone. Bulldozers and hand crews constructed fire lines and fuel breaks to protect the town. Large capacity sprinklers were installed on the southern and eastern sides of West Yellowstone and Old Faithful Village.

On September 6, winds picked up once again and by evening the fire had advanced to within one mile of Old Faithful Village. Strong winds of up to 50 mph on the 7th advanced the fire on all fronts and evacuation was ordered. As the fire swept through the area, it damaged 19 cabins, two dorm rooms, three storage buildings, a restroom, five vehicles, one water tank, and a television transmitter tower, but spared the Old Faithful Lodge. The fire was contained by mid-October, after burning a 385,035 acre area which included 372,199 acres within the park itself.

The "49er" Fire in California Gold Rush Country

In California, high winds fanned 11 wildfires through five rural counties in the Gold Rush country of the Sierra Nevada foothills in mid-September 1988. One of the largest, dubbed the "49er fire," was begun on September 11 by a vagrant burning toilet paper. The 49er fire skipped through a hilly and heavily forested section of northern Nevada County along Highway 49. In six days, the blaze forced closing of the highway, scorched about 33,500 acres of timber and brush, and destroyed homes and other buildings valued at 22,728,701 million dollars, according to Bob Paulus of the California Department of Forestry and Fire Protection, the ranger in the area affected by the fire.

In the affluent Lake Wildwood subdivision where million-dollar houses are not uncommon, 4,500 residents were asked to evacuate twice in two days. Residents of the nearby towns of Smartville, Rough and Ready, and Penn Valley were also evacuated. Firefighters were able to save most of the expensive Lake Wildwood homes, but in the end the blaze claimed 13 houses there, along with five boats and a dock.

At its height, 2,894 firefighters were involved in fighting the 49er fire, and the firefighting costs have been estimated in the neighborhood of 5.6 million dollars.

POLICIES AND ISSUES

Federal Fire Management Responsibilities

Wildland fire management policy is primarily an issue of land and resource management, and falls under the jurisdiction of the land management agencies responsible for national forests, parks, wilderness areas, and other public lands. There are five Federal land management agencies with wildfire management responsibilities:

- USFS

- NPS

- Bureau of Land Management (BLM)

- U.S. Fish and Wildlife Service (FWS)

- Bureau of Indian Affairs (BIA)

The USFS is an agency of the Department of Agriculture. The other four agencies are within the Department of the Interior. In addition, State agencies, usually within departments of forestry or natural resources, generally are responsible for wildfire management on State- and privately-held lands. (See Figure 1, "Wildfire Management: Federal and Non-Federal Agency Responsibilities.")

On the Federal level, the USFS is recognized as the leader in wildland fire management because of its legislative mandate, experience, resources, and expertise. The policies, and to a certain extent, the practices of the other four agencies closely follow those of the USFS.

There are some significant differences in the policies and practices of the NPS from the overall fire management direction provided by the Forest Service. NPS's mission is fundamentally different from the other four agencies. The missions and objectives of the USFS, FWS, BLM, and even the BIA are based on multiple use principles that balance protection of land with use and orderly development of natural resources. The NPS, however, has a single use mission: to preserve and protect its lands for the enjoyment of the American people. This fundamental difference in mission has had, and no doubt will continue to have, major implications for wildfire management.

Wildfire Management
Federal and Non-Federal Agency Responsibilities

Agency	Mission	Areas of Initial Attack Responsibility
Federal		
U.S. Forest Service (U.S. Department of Agriculture)	The U.S. Forest Service mission is to provide a continuing flow of natural resource goods and services to help meet the needs of the nation and to contribute to the needs of the International community. To accomplish this, the Forest Service administers the National Forest System under the principles of multiple use and sustained yield, according to the following objectives: provide a sustained flow of renewable resources -- outdoor recreation, forage, wood, water, wilderness, wildlife, and fish -- in a combination that best meets the needs of society now and in the future; administer the nonrenewable resources of the National Forest System to help meet the Nation's needs for energy and mineral resources; promote a healthy and productive environment for the nation's forests and rangelands; develop and make available scientific and technological capabilities to advance renewable natural resource conservation through cooperation with other Federal agencies and state and local governments.	The National Forest System of 156 national forests, 19 national grasslands, and 17 land utilization projects in 44 states, Puerto Rico, and territories. Approximately 32 million acres are set aside as wilderness and 175 million acres as primitive areas where timber will not be harvested.
National Park Service (U.S. Department of Interior)	The National Park Service administers for the American people an extensive system of national parks, monuments, historic sites and recreation areas. Its objectives are to administer the properties under its jurisdiction for the enjoyment and education of our citizens, to protect the natural environment of the areas, and to assist states, local governments, and citizen groups in the development of park areas, the protection of the natural environments, and the preservation of historic properties.	National Park and Recreation areas.
Bureau of Land Management (U.S. Department of Interior)	Responsible for the total management of 270 million acres of public lands and for subsurface resource management of an additional 300 million acres where mineral rights are owned by the Federal government. Resources managed include timber, hard minerals, oil and gas, geothermal energy, wildlife habitat, endangered plant and animal species, rangeland vegetation, recreation and cultural values, wild and scenic rivers, designated conservation and wilderness areas, and open space. Bureau programs provide for the protection (including fire suppression), orderly development, and use of public lands and resources under principles of multiple use and sustained yield. Land use plans are developed with public involvement to provide orderly use and development while maintaining and enhancing the quality of the environment. The Bureau also operates the Boise Interagency Fire Center (BIFC).	270 million acres of public lands located primarily in the Far West and Alaska and scattered parcels located in other states and an additional 300 million acres where mineral rights are owned by the Federal Government.
U.S. Fish and Wildlife Service (U.S. Department of Interior)	The U.S. Fish and Wildlife Service is responsible for the conservation, protection and enhancement of fish and wildlife and their habitats for the continuing benefit of the American people. In the area of resource management, the Service provides leadership for the protection and improvement of land and water environments which directly benefits the living natural resources and adds quality to human life. The Service's federal aid program apportions funds to the states for projects designed to conserve, develop and enhance the nation's fish and wildlife resources.	U.S. Fish and Wildlife Refuges.

Wildfire Management
Federal and Non-Federal Agency Responsibilities

Agency	Mission	Areas of Initial Attack Responsibility
Bureau of Indian Affairs (U.S. Department of Interior)	The principal objectives of the Bureau are to actively encourage and train Indian and Alaska Native people to manage their own affairs under the trust relationship to the Federal Government; to facilitate, with maximum involvement of Indian and Alaska Native people, full development of their human and natural resources potential; and to mobilize all public and private aids to the advancement of Indian and Alaska Native people in the direction and management of programs for their benefit. The Bureau acts as trustee for Indian and Alaska Native lands and monies held in trust by the United States, assisting them to realize maximum benefits from such resources.	Individual and tribal lands held in trust for Indian and Alaska Native people.
Non-Federal		
State land management agencies, usually state departments of forestry or natural resources.	Varies.	State- and privately-held wildlands. Varies by state. Most states set burn policies, conduct surveillance, and most fire suppression actions even on privately-held lands. There are a number of cost-sharing schemes that require industrial and non-industrial forest owners to pay assessments for this protection into a state firefighting fund. In Oregon and Washington, however, the industrial owner has legal responsibility for fire protection. Some industrial owners maintain firefighting capabilities, but most pay the state to provide the service. In California, the state has the legal responsibility for fire protection on privately-held lands.
Local fire management districts, local paid and volunteer fire departments.	Varies	Varies. Sometimes responsible for protection of structures on non-industrial privately-held wildlands, while state agencies have responsibility for wildland protection.
Private foresters/ industrial owners	Not applicable.	Varies. In some states, private foresters have legal responsibility for fire protection in their forests. In other states, the state has legal responsibility. In either case, formal agreement and cost sharing principals determine actual policies. In Oregon and Washington, for example, the industrial owner has legal responsibility for fire protection. Some industrial owners maintain firefighting capabilities, but most pay the state to provide the service. In California, the state has the legal responsibility for fire protection on privately-held lands. There are a number of cost-sharing schemes that require industrial and non-industrial forest owners to pay assessments for this protection into a state firefighting fund.

SOURCES: U.S. Governmental Manual
Interviews with spokesmen for the U.S. Forest Service
Bureau of Land Management, Bureau of Indian Affairs, and National Park Service

1879-1:25 91-2

Development of Federal Fire Management Policies

The first wildland fire control program was established in 1885 in the Adirondacks Reserve in New York.[2] By the following year, a program was established in the West in Yellowstone National Park. Both were based on practices in use in Germany, then considered the world's model for forestry, and upon which early U.S. foresters modeled many forest management policies. The German practice was to put out all fires, and that practice was adopted in the United States.

In a pattern that would be repeated several times in intervening years and again following the 1988 season, devastating fires provided the catalyst for the first significant re-examination of U.S. wildland fire policy. In 1910, after catastrophic blazes burned more than five million acres and killed 79 firefighters in the Southern Rockies, the USFS adopted a clear, single fire management philosophy that declared "the moral equivalent of war" on forest fires. This philosophy was to guide its policy for the next 50 years.

The policy was simple: stress fire prevention and control all fires as quickly as possible. Fire was seen as a disruption of natural processes in the wilderness and should be stopped. During this time, U.S. foresters saw their jobs largely in terms of stopping fires, and success in fire management was measured by how many fires were extinguished. The area with the fewest fires was considered the best managed.

In the early 1920s, there was some debate over the merits of light burning versus a policy of total suppression, but control of forest fires continued to be the paramount concern within the Forest Service. During this period, a 10-acre control objective that allowed some burning to occur but called for suppression of all fires more than ten acres, was adopted in 1926. The revised Forest Service policy also reflected a new economic philosophy which assumed that the sum of suppression cost plus loss would be less under an all out suppression policy that eliminated uncertainty and lack of aggressiveness.

The Tillamook Burn and the "10 a.m. Policy"

The Tillamook burn of 1933, which destroyed three million acres of essentially virgin timberland in the Northwest, had a profound effect on Forest Service fire policy. The experience led the USFS to adopt an even more rigid form of its basic "no burn" policy. All fires were to be controlled as quickly as possible, preferably during the first duty shift after detection. If that wasn't possible, fires were to be controlled by 10 a.m. of the following day.

This "10 a.m." policy was made possible with the availability of labor through the newly-established Civilian Conversation Corps (CCC) and with the emergence of organized emergency fire crews, the management of organized fire suppression forces, and the development of formal line construction methods. The 10 a.m. policy was consistent with the cost-plus-loss economic-based objectives established in 1926. This policy survived until the early 1970s.

---

[2] Information for this section comes primarily from *Fire in American* by historian Stephen J. Pyne and from written testimony and background materials that reference the Pyne book provided by Interior Secretary Donald P. Hodel and George S. Dunlop, Assistant Secretary of Special Services, United States Department of Agriculture, at a September 29, 1988 hearing before two Senate subcommittees. Where information was gathered through personal interviews, it is attributed to its source. Other sources include several newspaper stories and an article that appeared in the September issue of "Roots," published by the Minnesota Department of natural Resources.

The "Leopold Report": The Role of Natural Fire and Use of Prescribed Burns in the Park Service

Two events – the issuance of the "Leopold Report" in 1963 and the subsequent enactment of the "Wilderness Act of 1964" – challenged the principles which had guided U.S. wildland fire policy since its beginnings, led to a new view of the role of natural fire in wilderness and inaugurated a slow transformation of fire policy.

Says historian Stephen Pyne, "Fire control in itself was now considered inadequate – indeed ruinous – as a program of resource management." The Leopold Report stressed the importance of letting fire play its natural role in the ecosystems and cited experiments in controlled burning in the Everglades. It recommended the controlled use of fire as one of the most natural, as well as least expensive and most easily applied methods of manipulating vegetation in wilderness areas, and urged that the Park Service institute a research program to guide its resource management objectives. Soon thereafter the National Academy of Sciences also recommended that the Park Service institute a research branch.

One of the first NPS studies conducted looked at the effects of fire on the giant sequoia tree and challenged conventional thinking. For years, conservationists had removed debris in sequoia groves by hand, even though in earlier times Indians and loggers had routinely burned slash without real damage to the huge trees. One study focused on the relationship of fire to sequoia regeneration, while the second was a survey of fuel hazards around sequoia groves. Both refuted the popular belief that fire was harmful to the sequoias and led to recommendations for use of prescribed burning.

In 1968, the NPS issued a controversial new policy handbook (NPS-18) that recognized fire as a natural process, encouraged letting natural fires run their course in predetermined areas, allowed the use of prescribed fire as a substitute for natural fire where necessary, and called for control of any fire that did not further management goals. It would be nearly ten years before the USFS would implement a similar policy allowing for use of prescribed fire as a land management tool.

Following introduction of the new policy handbook, Pyne writes, the Park Service experimented with fire, both through research and in its practices. While the handbook provided some guidance to the application of Park Service policies, for the next ten years the extreme flexibility of the policy allowed for variety and adaptability to local conditions. Pyne argues that this flexibility also left NPS fire policy somewhat fragmented and unduly influenced by local "peculiarities."

Adverse public opinion led the NPS to take another look at its fire policies in the mid 1970s, when residents of Jackson Hole, Wyoming, were angered by thick smoke from a smoldering natural fire in the Tetons. The Park Service responded by issuing a set of interim guidelines that provided more specific instructions for conducting its fire program. In 1976, a task force was charged with re-evaluating all NPS resource management plans, and a detailed handbook (NPS-18) was issued in 1978, thus completing a ten-year implementation process that began with introduction of the Park Service fire management policy handbook in 1968.

Natural Fire and Use of Prescribed Burns by the Forest Service

While Park Service fire management policies changed dramatically during the 1960s, between 1935 and the early 1970s, the Forest Service basically maintained its "no burn" policy and USFS fire management policy was the same for both wilderness and non-wilderness areas.

In 1971, the Forest Service re-examined its "10 a.m. policy," primarily because of three concerns. First, there was a concern that the long history of prompt suppression had caused an increasing buildup of fuels in wilderness areas. Second, the 10 a.m. policy was not allowing lightning-caused fires to play their natural role. Third, the quick suppression policy was too expensive and did not allow forest managers the flexibility to weigh the cost of taking immediate suppression action against the resource values at risk.

As a result of the 1971 re-examination, the Forest Service adopted a major modification of its policy by allowing lightning-caused fires to burn under specified conditions in Congressionally-designated wilderness areas having an approved fire management plan. These were designated natural "pre-scribed fires." All other fires, including all human-caused fires, were considered wildfires and were to be immediately suppressed. The 10-acre suppression policy was incorporated as a pre-suppression planning objective.[3]

In response to the 1971 policy change, a major fire planning effort was undertaken. As a result, Forest Service costs skyrocketed in the 1970s for pre-suppression activities. Numbers of fires and acres burned also increased in spite of expanded pre-suppression programs in many of the wildland fire control agencies.

The Forest and Rangeland Renewable Resources Planning Act of 1974 and the National Forest Management Act of 1976 required that both the use of prescribed fire and the control of wildfire be an integral part of the Forest Service land management planning process. Further, fire management was to be responsive to resource management objectives in a cost-effective manner. Thus, in 1977, the Forest Service directed that a study of pre-suppression effectiveness be conducted. The study concluded that fire management objectives must be directly related to resource values and to the costs of protecting them, and that protection should be commensurate with values and risks to that the program is accountable, efficient, and cost-effective.

In 1978, the Forest Service again made a significant revision in its fire management policy. The objective of wildfire suppression was changed from one of control of all wildfires by 10 a.m. to one of managing fire suppression costs and damages, consistent with land and resource management objectives. Prescribed fire to protect, maintain, and enhance national forest resources was reaffirmed as an approved management practice.

Common Ground Is Reached

By 1978, both the NPS and the USFS reached a point where their policies institutionalized a common philosophy about the natural role of fire and the use of prescribed burns, albeit their missions dictated practices that implemented them directly.

Between then and 1988, there were no major revisions in Forest Service or Park Service fire management policies. In 1985, the Forest Service issued a clarification of its objectives for the use and management of naturally-ignited and human-caused fires in Congressionally-designated wilderness areas. Forest Service policy allowed natural fires to burn when necessary to meet the objectives of (1) allowing these fires to play their natural ecological role and (2) reducing the risks of wildfire

---

[3] Pre-suppression refers to activities undertaken prior to a fire's designation as a wildfire. Pre-suppression activities include all risk reduction measures ranging from the development of fire plans, storing of firefighting equipment and supplies, and reduction of hazard fuels through the use of prescribed burns. All pre-suppression activities must be budgeted, and emergency firefighting dollars that fund wildfire suppression efforts are not available for these efforts. Forest Service and Park Service policies distinguish between suppression and pre-suppression activities to separate these expenditures.

and to life and property within wilderness, and to life, property, and resources outside of wilderness areas.

The National Park Service's NPS-18 was already under revision when the 1988 wildfire season began. That process continued in the wake of the policy debate that followed, and new guidance was issued early in 1991.

The Role of the Natural Fire in Wilderness Areas

As noted earlier, NPS fire management policy reflects a basic belief that fire is natural to many wildlands and plays a vital role in the ecology of those lands. Rather than destroy the lands, fire actually aids in the perpetuation of some species of plants, including the lodgepole pines that dominate Yellowstone. Their seeds are not released unless the intense heat of fire opens the pine cones. Other species, such as the Ponderosa pine, have developed a thick bark that protects them against fire.

Fire historian Stephen Pyne of Arizona State University explained in a *New York Times* interview that fire aids regeneration by eliminating a naturally occurring chemical that inhibits excessive budding in healthy plants. It also seems to release a number of minerals and chemicals that otherwise would not be released.

The Park Service's belief in the importance of natural fire to ecosystems such as Yellowstone is one shared by many conservationists. The Wilderness Society's Barry Flamm, a former chief of fire management for the Forest Service, points out that fire has played a critical role in the evolution of the park. Said Flamm in a 1988 interview, "These fires are part of a natural cycle, and in the spring we expect to see the regeneration that always follows a fire. It happens quickly, and the changes benefit many species."

That view was reiterated by a team of experts, chaired by Norman L. Christensen of Duke University, who met under the sponsorship of the Greater Yellowstone Coordinating Committee in November 1988 and January 1989 to assess the ecological effects of the Greater Yellowstone Area fires. The final report of the Greater Yellowstone Postfire Ecological Assessment Workshop states:

> Rather than ecological disasters or catastrophes, high-intensity fires in ecosystems such as those of the GYA are virtually inevitable, and are even essential for the successful reproduction of some species.[4]

In fact, regeneration had already begun in the weeks just following the fires, according to Yellowstone spokesperson Sandy Robinson. She reported that by mid-October 1988, grasses and other small plants were already six inches to a foot high in some burned areas. Rather than vast areas of charred landscape, the area looked more like a mosaic, with no burned area too far from another that still had vegetation providing a nearby source of seeds and wildlife for regeneration.

Nor do the fires destroy everything in their paths. They often flash through an area, burning brush, grass and fallen trees, but only searing healthy trees. Of the estimated 793,880 acres of the 2.2 million acre park reported burned, 323,291 acres sustained canopy burn, 281,098 acres sustained a combination of canopy and surface burn, 51,301 acres of meadow, grassland or sagebrush burned, and another 138,190 acres sustained burns to the edges of forest areas or to isolated stands of trees in unforested areas.

Before the fire, many parts of Yellowstone were dominated by large areas of old lodgepole pines and large blocks of vegetation limited to certain species locked in over time. Many were diseased with infestations of the pine bark beetle and dwarf mistletoe, which next to fire, are the major killers of the trees. The Park Service and many conservationists do not consider such homogeneous blocks of vegetation natural or desirable, and many say their development was facilitated by the earlier "no fire" policy that did not allow fire to play its "clean out" role through the years.

The long-standing "no fire" rule may also have helped create the huge accumulation of fuels that contributed to the ferocity of the Yellowstone fires. Not only were many over-age trees still standing, but many fallen trees lay in jackstrawed piles, providing a further source of fuel. A scientist at the Forest Service's Intermountain Fire Sciences Laboratory in Missoula, Montana, estimates that in 1988 every acre of Yellowstone contained 40 to 45 tons of small dried matter. Just one Douglas fir tree with a 30-inch diameter contributes 17,000 pounds of flammable wood fibers.

However, the Greater Yellowstone Postfire Assessment Workshop concluded that fuel load was only one factor contributing to the ferocity of the 1988 conflagration, and may not have been the dominant factor. The final report states:

> The large-scale behavior and extent of the fires appear to have been established more by drought and wind, although fuel distribution certainly affected fire intensity and behavior on a local scale. The events of 1988 demonstrated that fire suppression in heavy fuels may be impossible when weather is severe.[4]

With large stands of mature trees downed by the fire, new plants and species will now have the necessary space and sunlight to grow, and scientists predict that areas with more diverse vegetation will eventually replace most, but not all, of the areas burned this year.

Christensen and his colleagues predicted that as a result of the 1988 fires, meadows and young forests will cover larger portions of the Greater Yellowstone area for decades. Aspen forests will occupy a larger (albeit still small) portion of the area, and some areas will have less sagebrush for at least the next decade. Other parts of the park and surrounding areas will be dominated by more luxuriant herbaceous vegetation for several decades, while still others will remain barren for years to come.

Thus, while many conservationists believe that the huge Yellowstone area fires were ecologically beneficial, it could be generations before Yellowstone is restored to its former grandeur.

A Comparison: Canada's Wildland Fire Policy

While the NPS and the USFS "let burn" policies were criticized for contributing to the huge fire losses in 1988, Canada lost less acreage in its national parks and forests than it had in decades. Some point to the Canadian policy calling for suppression of 85 to 90 percent of forest fires as soon as they are detected as a major factor contributing to its strikingly different experience that year, but even Canadian forest officials said that they were just plain lucky. The same weather pattern that caused the severe drought in the U.S. brought higher than average rainfall to the parts of Canada that contain large areas of park and forest land. The southern areas affected by the drought had late winter snowfalls that helped to mitigate its effects.

According to David Lohnes, Director of Natural Resources for the Canadian Park Service, Canada's policy of immediately suppressing most forest fires is not based on a disagreement with the U.S. view of the role of natural fire. In a *New York Times* story, Lohnes explained that Canada fights almost all

---

[4] Ibid., p. iv.

of the fires in its national parks because of their proximity to the valuable timberlands that support the country's large lumber industry. In fact, after the U.S. adopted its so-called "let burn" policy, Canadians debated adopting a similar position. In 1979 they modified their approach by allowing some safe and ecologically helpful fires to burn.

Lohnes, who favors a more liberalized policy allowing natural fires to burn in Canada's park lands, said that American forest management is considered the most advanced in the world. He pointed out that despite its policy of suppressing most natural fires, Canada too, has experienced devastating natural fires. The worst year was 1981, when a total of 1.7 million acres of Canada's park land was consumed by fire, including 238,000 in its largest park, the Wood Buffalo National Park.

The "Let Burn" Policy And The Use Of Prescribed Fire In Wildlands In 1988

As the fires raged through Yellowstone in 1988, there was widespread public misperception of the NPS fire management policy, which is often referred to as its "let burn" policy. In reality, park staff evaluates every fire that occurs, Interior Secretary Donald P. Hodel told two Congressional committees in testimony September 29 of that year.

That continues to be true today. In fact, following the major re-examination of fire management policies after the 1988 wildfire season, NPS and USFS fire management policies were reaffirmed and strengthened. What have changed are guidelines and procedures spelling out how those policies are implemented.

Local Fire Management Plans

Approved plans encompassing all land management objectives are required for each national park and forest. Fire management plans are operational contingency plans, constitute an important component of each area's comprehensive land management plan, and are formulated at the local level subject to public comment and review.[5]

The planning process begins with an assessment of public concerns through public hearings in and around the national park or forest and in communities that are dependent on these lands for economic, recreational, social, or other reasons. Public comments are formally sought and are then analyzed before a final plan is published. Once again, the public has an opportunity to appeal the final plan through a formal appeals process.

These comprehensive land management plans are updated periodically as needed through an amendments process, and must be revised every 10-15 years. While the amendment process is initiated within the Park Service or Forest Service itself, the need for amendment can be triggered by calls for such action from surrounding communities as well as by major changes in resources, physical, social or economic conditions, or other factors. Because fire affects these factors so significantly, amendments are often initiated following a major fire.

The Yellowstone Plan was developed in 1972 and updated in 1976. Like all national park management plans, it was subject to public comment and review. Ironically, the Yellowstone Plan had been updated again in 1988 but the new one had not yet been formally approved at the time of the fires.

---

[5] See page 34 for information about how to investigate opportunities for involvement in this process.

Prescribed Fire Plans

Both the NPS and the Forest Service require a detailed fire management plan for each area in which use of prescribed fire is authorized. The prescribed fire management plan provides a specific "prescription" that, for each fire, specifies the area in which it will be allowed to burn; the weather conditions under which it will be allowed to burn; and a number of "contingency levels" which are basically decision trees that show how strategies would be affected by changes in fire behavior, weather conditions, availability of firefighting resources and other variables.

When a prescribed natural fire exceeds its prescription, it is reclassified as a wildfire. The primary criteria for reclassification are threats to life and property, ability to keep the fire within the park/forest boundary and resources available to control the fire. If any of these change, or if weather conditions or other factors change, the prescribed natural fire is reclassified as a wildfire and the appropriate suppression action begins.

Although complex models were used to set prescriptions by predicting fire behavior under various conditions, these models were found to be inadequate and over-reliant on historical data on weather patterns and fire behavior. In 1988, neither the Park Service nor the Forest Service explicitly recognized the inherent uncertainty in setting the parameters in a fire prescription. That is, the models used to predict behavior or prescribe parameters did not include numerical measures of uncertainty (the probability of exceeding predictions) that defined the inherent limitations of the models as predictors.

These shortcomings meant that those fighting the blazes often failed to take into account several factors critical to making good firefighting decisions.

Detection and Suppression

According to Steve Hodap of the NPS, all fires that occur on Park Service lands are closely monitored. The method of monitoring varies from area to area depending on the value of resources involved and the area's proximity to people, structures, and other resources. Routine monitoring may include flyovers as well as close monitoring by highly trained fire observers. Routine flights are made after lightning storms to detect small natural fires, and those detected are monitored regularly. Late detection is rarely a factor in a fire's becoming a wildfire, according to Hodap. "We are aware of virtually every significant fire that starts on Park Service lands within 24 hours," he said.

On lands protected by the USFS, about 90 percent of the time "appropriate suppression action" means immediate suppression, but it can also mean a decision to simply monitor a fire that is expected to extinguish itself or contain itself in rugged terrain. Suppression action may be taken later, if it appears that the fire is not behaving as originally anticipated.

If a fire escapes initial attack, an Escaped Fire Situation Analysis (EFSA) is prepared. This is a written document that assesses the potential suppression costs versus potential resource losses, threats to life and property in the short and long run, firefighter safety factors and other factors to set objectives for fighting the fire. Generally, this analysis is prepared by a line officer at the forest's district or supervisory level.

A series of incident briefings are held and the firefighting objectives in the EFSA are translated into specific operational strategies by the incident management teams that fight the fires. Depending on the size of the blaze, these teams can be organized at the local, district, forest or regional levels. For the largest fires, interagency teams dispatched through the Boise Interagency Fire Center will manage the firefighting efforts.

A similar process is followed in the NPS, but some of the factors that go into the equation are different and produce variations in the application of fire management policies. Park Service policy dictates that "within the framework of management objectives and plans, overall wildland fire damage will be held to the minimum possible giving full consideration to: (1) an aggressive fire prevention program; (2) the least expenditure of public funds for effective suppression; (3) the methods of suppression least damaging to resources and the environment; and (4) the integration of cooperative actions by agencies of the Department of the Interior among themselves or with other qualified suppression organizations."

Interagency and Intergovernmental Coordination of Wildland Firefighting Efforts

As wildland fire protection programs became larger and more sophisticated, Federal and State agencies found it necessary to formally coordinate programs, standards, and procedures. The NWCG was chartered in 1976 for this purpose.

The National Interagency Fire Qualification System (NIFQS) was developed under the sponsorship of the NWCG for the purpose of assuring a nationwide source of professional wildland firefighting personnel.

The increasing complexity of wildland fire suppression in the urban/wildland interface in southern California and a terrible 1970 fire season led to the implementation of California's FIRESCOPE (**Fi**refighting **Re**sources of **S**outhern **C**alifornia **O**rganized against **P**otential **E**mergencies) project, where the Federal wildland fire service, the State of California, and various city and county jurisdictions officially work together as a single team to fight wildland/urban interface problems. The Large Fire Organization (LFO) and Incident Command System (ICS) evolved from this program in response to the need to more effectively integrate the suppression resources of Federal, State, and local fire protection agencies.

In 1980, the NWCG sponsored a study to evaluate the LFO and the ICS. As a result, NWCG adopted the National Interagency Incident Management System (NIIMS). NIIMS built on the strengths of the LFO, ICS, and other FIRESCOPE technologies to provide a common, integrated emergency management system for the interagency management of emergency incidents of all types. By 1982, all Federal land management agencies with wildfire management responsibilities and many States had adopted the NIIMS for implementation. The ICS element of NIIMS was fully implemented by all Federal agencies and in most States by 1985.

Implementation Of Federal Policies In The Yellowstone Fires

The Greater Yellowstone Area Fires provided a crucial test of the implementation of NPS and Forest Service wildfire management policies under extreme conditions. While many critics have deplored NPS policies that some say over-emphasized the role of natural fire and precluded use of the most aggressive suppression actions, other argued that the conditions under which the fires burned were so extraordinary that they could not have been suppressed anyway.

That was the conclusion reached by the academicians who participated in the Greater Yellowstone Postfire Ecological Workshop and the government officials who served as members of the Interagency Fire Management Policy Review Team.

Implementation of the "let burn" policy: Of the 13 major fires that burned in Yellowstone National Park in 1988, eight were fought from the start. Only the earliest lightning-ignited fires in the park were allowed initially to burn, in accordance with the NPS's prescribed natural fire policy. In his September 29 statement to two Congressional committees, Secretary Hodel said that the "let burn" decision was based on the higher than average spring rainfall in the park. When rainfall fell well below normal later in the season, the Park Service decided to make an exception to the policy and began all out suppression efforts. By then, it was already too late.

According to the NPS's "Wildland Fire Report: 1988," unprecedented fire behavior and the severe weather conditions rendered most accepted firefighting techniques virtually useless. Fire lines constructed along the edges of the advancing fires to create fuel breaks and backfires to reduce fuel accumulations in front of advancing fires were frequently ineffective because fires spread long distances by "spotting," a phenomenon by which wind carried embers from the tops of the 200-foot flames far out across unburned forest to start spot fires well ahead of the main fire. Spotting often occurred up to a mile and a half ahead of the main fire. The blazes regularly jumped the widest bulldozer lines and even jumped traditionally recognized barriers such as the Grand Canyon of the Yellowstone River.

The fires often moved two miles an hour and advanced up to five or ten miles per day, consuming light fuels that would have been unburnable in an average season. Drought conditions were so severe that fires could not even be fought at night, when they normally "lie down" as increased humidity and lower temperatures quiet them.

By July 15, the NPS report states, "The experts on site generally agreed that without help from the weather, in the form of rain or snow, there was no technology in existence that could stop the fires." And indeed, it was not until mid-September, when a series of cold fronts began moving through the Northwest, that cooler temperatures and precipitation provided the break that allowed firefighters to attack and eventually contain the fires.

NPS and USFS Fire Investigation Reports

On December 2, the NPS and the USFS issued detailed fire investigation reports on the four largest Greater Yellowstone Area fires, including the North Fork and the Clover-Mist blazes. The reports concluded that the "let burn" policy ought to be retained, but were critical of the Park Service's implementation of the policy. This early conclusion was later reiterated by two separate panels of experts charged with reviewing NPS/USFS policies following the 1988 fires: the Greater Yellowstone Postfire Ecological Workshop and the Interagency Fire Management Policy Review Team.

The NPS/USFS report on the Clover-Mist fire says that it could have been suppressed in early July if park managers had decided to fight the fire early. Instead, they waited until July 21 to begin suppression action, because park officials relied heavily on historical fire data when they made their critical early decisions not to fight the fires. Despite unprecedented drought conditions, they continued to believe that the fire would not exceed their "worst case scenario" of 40,000 acres. When it was over, the Clover-Mist fire had grown to almost ten times that size.

Implementation of the use of firefighting techniques and equipment that are "least damaging to resources and the environment": According to the NPS, about 665 miles of handlines and 137 miles of bulldozer lines were constructed to fight the GYA fires, including 32 miles of bulldozer line in Yellowstone Park itself. About 1.4 million gallons of fire retardant were dropped, 10 million gallons of water were dropped by helicopters alone, and innumerable water pumping stations were established. There were also 51 spike camps, 150 helispots and a major camp established for each fire. There were significant short term impacts to wilderness and grizzly bear habitat because of the use of motorized vehicles.

Still, firefighter battling the Greater Yellowstone fires were sometimes frustrated by the Park Service's reluctance to allow them to use techniques and heavy equipment that they believed necessary to their efforts, according to accounts in the *New York Times* and *Time* magazine. One crew was reportedly reproached for crossing a meadow with their fire truck to put out a spot fire, while a number of helicopter pilots said that Park Service restrictions on where they could dip their massive buckets made their job more difficult. Even at the height of the fires, the *New York Times* reported, bulldozers were allowed in the Park only on a case-by-case basis.

Heavy equipment such as bulldozers to make fire lines and clear brush were not used to fight the huge Clover-Mist fire. While Interior Secretary Hodel had ordered its suppression in mid-July, a park spokesperson acknowledged that the portion of the Clover Mist fire within park boundaries was not fought until August 9 because officials thought natural barriers such as cliffs and streams would stop its spread.

Fire Management Policy Review Team Recommendations

In late September 1988, Interior Secretary Donald Hodel and Agriculture Secretary Richard E. Lyng announced formation of the Fire Management Policy Review Team to review fire management policies and their application for national parks and forests and to recommend actions addressing the problems experienced during the 1988 fire season.

The Fire Management Policy Review Team included representatives from the five land management agencies with wildfire management responsibilities – the NPS, BLM, FWS, and BIA in the Interior Department, the USFS in the Department of Agriculture, the Boise Interagency Fire Center and the National Association of State Foresters.

The panel consulted knowledgeable organizations and individuals including governors, local government officials, concessionaires and outfitters, individuals with businesses in nearby communities and organizations with an interest in parks and wilderness areas. The review team met regularly with representatives of the National Fire Protection Association (NFPA), the Western Governors Association, and the academic community.

The team issued its draft report on December 15, 1988. Following a 60-day period for public comment, the team's final report was submitted on May 5, 1989. The team concluded that the basic objectives of Federal fire management policies were sound but needed to be refined, strengthened, and implemented correctly. Many of the plans devised for individual forests, parks, and wilderness areas did not conform to policy guidelines. The panel also found serious shortcomings in many of the fire models or "prescriptions" specifying in what areas and under what conditions natural or ignited fires will be allowed to burn.

Among the major findings of the Interagency Fire Management Policy Review Team were the following:

The "Prescribed Natural Fire" Policy

The policy governing use of prescribed natural fire was sometimes abused. Plans were developed that did not meet the literal requirements of the policy, and some plans did not contain adequate prescription criteria. Many of the wilderness and park land managers had such a strong belief in the beneficial role of natural fire that they applied less stringent controls over prescribed fires than was required by policy.

The team cautioned against exclusive focus on the ecological benefits of natural fire, which they said can lead to inadequate consideration of the impacts of fires on recreation, wildlife habitat, grazing, and water quality. They also reported hearing from agency employees who wanted the current policies to be expanded to allow fires to burn without prescription as long as they are not expected to cross the administrative boundaries of a park or wilderness area or endanger human life and property.

Inadequacy of Many Fire Management Plans

The panel also found that "fire management programs would be strengthened by a combination of improved decision criteria in plans, additional fire expertise, and more direct line officer involvement." They cited a lack of identified critical decision points in many fire plans, as well as a "critical lack of resident fire expertise in some locations." They also pointed out that some fire management plans did not include the latest technology and were not complete in terms of indicators of long-term drought and impact on shared suppression resources.

Some plans did not sufficiently address suppression resource availability, values at risk outside of parks and wilderness and the number of fires that can be managed at one time. Also, some models did not address cumulative effects of drought and other potentially important considerations.

The review team noted that, as a result of these shortcomings, some fires eventually declared wildfires continued to be considered as prescribed fires until they reached an arbitrary limitation such as a park or wilderness boundary. Insufficient attention was given to values at risk both inside and outside parks and wilderness boundaries, as well as to the cumulative risks associated with multiple fires, large fires or fires with especially active perimeters.

Lack of research and analysis upon which to base fire behavior predictions included in models: The team found a need for research and analysis to provide tools for management of fire management programs. They noted that while normal climatic patterns are ordinarily used for projections, prolonged drought periods may affect these patterns and hamper ability to project fire behavior accurately. Similarly, the team found that analyses of fire history, occurrence, size, and effects were insufficient for many areas.

Underutilization of Ignited Prescribed Fires to Reduce Accumulated Fuels

The team noted that the reduction of hazard fuels around structural developments, parks/wilderness boundaries and private holdings if often desirable because it enhances the ability to protect these areas and reduces costs of wildfire suppression and prescribed natural fire. However, they found that the use of planned ignitions is too limited in some areas.

Inadequacy of agency personnel development and training programs: The team found too few professional managers in field locations with an understanding of fire management and fire management policies and practices. They found that many line officers did not require adherence to standards in fire management plans and some incident management teams, fire professionals, and line officers "lack knowledge of suppression tactics necessary under extreme conditions." The team also noted that some agency fire staffs do not maintain expertise in fire management because of infrequent fires and the lack of career mobility and opportunities to gain experience in other locations.

Based on these and other findings, the team recommended the following:

- fire management plans be strengthened by:

    (a) developing joint agency fire management plans, agreements, or addenda to existing plans for those areas where fires could cross administrative boundaries;

    (b) including a comprehensive set of new criteria to be used in deciding whether or not to allow natural ignitions to burn as prescribed fires;[6]

    (c) clearly describing the decision process and factors to be addressed before a fire is declared a prescribed natural fire;

    (d) including criteria (such as consideration of the effects of cumulative drought conditions on fire behavior, availability of firefighting resources, and consideration of the regional and national wildfire situation) to be used in declaring a prescribed fire a wildfire;

    (e) clearly identifying areas that need protection from fire, such as developments within or adjacent to wilderness and park boundaries, and including within fire management plans actions that are to be taken to protect them, i.e., hazard fuel reduction or installing fuel breaks;

    (f) clearly stating the management objectives being addressed by the prescribed natural fire program; and,

    (g) clearly describing the process to be used to ensure adequate public involvement and coordination with local governments.

- regional and national contingency plans be developed to constrain prescribed fires under extreme conditions

- agencies consider planned ignitions to complement prescribed natural fire programs and to reduce hazard fuels

- more consideration be given to the effects of fire policies in surrounding communities and steps be taken to encourage more public participation in the development of individual fire management plans

- additional research and analysis on weather, fire behavior, fire history, and fire information integration be carried out so that future fire management programs can be more effective in reducing the risk of wildfires.

---

[6] Existing criteria already included: consideration of land management objectives for the area, historic fire behavior, natural role of fire, proposed degree of suppression, expected fire behavior, acceptable suppression techniques, adequate buffer zones, smoke management and effects on adjacent landowners. New criteria added include: consideration of the effects of cumulative drought conditions on fire behavior, availability of firefighting resources and consideration of the regional and national wildfire situation, to name a few.

The panel also cited wildland/urban interface issues as an area in need of further study. Specifically, it recommended further study to determine the current and future effects of residential and commercial development on the ability to design and implement prescribed fire programs and to examine the interrelationship between fire management plans and local planning and zoning functions.

Post-Yellowstone NPS and USFS Fire Management

Following the 1988 wildfire season, both the NPS and the USFS took steps to implement the recommendations of the Fire Management Policy Review Team. Among the steps taken were the following:

- a detailed review was immediately undertaken of all fire management plans in national parks and forests that included use of "prescribed natural fire" (naturally-ignited fires allowed to burn under certain "prescribed" conditions)

- both the NPS and the USFS imposed a temporary freeze on their "let burn" policies until fire management plans were revised to meet current policy and new requirements

- the NPS instituted a daily certification process requiring verification of the availability of adequate firefighting resources to assure prescribed natural fires will remain within prescriptions. If not, they are to be declared wildfires and suppression actions are to be taken immediately. NPS also reviewed its policies regarding the use of prescribed burning (e.g., planned ignitions) in place of prescribed natural fire

- the NWCG was ordered to develop criteria for a national contingency plan for controlling natural fires under extreme conditions, and regional offices at the National Park Service were ordered to develop regional contingency plans

- the USFS revised its fire management manual (Forest Service Manual 5100 – Fire Management) to implement the recommendations of its 1989 in-house Task Force Report on Prescribed Fire Management Criteria. This report enumerated detailed recommendations for the Forest Service's policy and guidance documents spelling out how policies are to be implemented

- the NPS issued a revised for management policy manual (NPS-18) in early 1991. This completes a revision process that began before the 1988 wildfire season and was expanded as a result of the fires.

The 1990 Fires

As noted earlier, there has been no real operational test of the post-Yellowstone fire management policies for national parks, forests, and wilderness areas although fires in two national parks during the 1990 wildfire season threatened, but ultimately failed to rekindle old controversies.

The 1990 wildfires that closed Yosemite National Park had all the drama, but little of the controversy that surrounded the policy debates during the Greater Yellowstone Area fires of 1988. Like the Yellowstone fires two years earlier, the fires in Yosemite and the adjoining Stanislaus National Forest were natural fires started by lightning strikes in a drought-dried area littered with decades of accumulated dried, dead wood. But this time there was no question that the blazes would be fought immediately, writes Larry B. Stammer in an August 11, 1990, *Los Angeles Times* article. These fires burned in "suppression areas," where lives and property were at risk.

The largest of the fires, named the A-Rock fire because of its proximity to the Arch Rock at the entrance of Yosemite National Park, burned more than 8,500 acres in the park and 11,000 acres in the Stanislaus National Forest. The blaze, whose 500-foot flames burned so intensely that granite boulders were fractured, destroyed the small summer cabin community of Foresta.

The other large fire in 1990 supplied a partial first test of the post-Yellowstone fire management policies. Ironically, the fire occurred in Kings Canyon National Park, which along with Sequoia National Park, was the first to reinstate the NPS policy of letting some natural fires burn themselves out. The ban on allowing natural fires to burn in the Kings Canyon and Sequoia Parks was lifted late in 1989 after NPS regional director Lew Albert approved revised fire management plans for the parks that allow natural fires to burn within a specific range of temperature, wind speed, and fuel moisture conditions.

As a result, the 1990 fire in the Kings Canyon National Park, located in remote high country away from people and property, was allowed to burn as a prescribed natural fire for three weeks during the summer. However, it was reclassified as a wildfire under the post-Yellowstone guidelines because the regional wildfire situation had become severe and firefighting resources were scarce, according to Doug Erskine, NPS fire director at the Boise Interagency Fire Center.

The post-Yellowstone guidelines call for preparation of interagency wildfire emergency preparedness plans that include careful consideration of the availability of firefighting resources on a regional and national basis. Once specific thresholds are reached, the plans specify that no new fires would be allowed to burn as prescribed fires. When firefighting resources were stretched even further (as was true in the Kings Canyon fire), a second threshold is reached and fires previously classified as prescribed fires are to be aggressively fought.

THE WILDLAND/URBAN INTERFACE FIRE PROBLEM

The wildland/urban interface refers to the area where relatively untouched wildlands and residential areas meet. At one time the distinction between the two was relatively clear, but changing residential patterns in recent years have seen increasing numbers of people building homes in and around wildland areas in all regions of the Nation. U.S. Census figures indicate that the population in rural areas grew faster than urban areas during the 1970s, although the trend has slowed somewhat since then. Population in areas near national forests grew 23.4 percent between 1970 and 1980, according to the USFS.

In some wildland/urban interface areas, isolated homes and other structures are scattered throughout a wildland area. Individual homeowners face a significant risk because of their isolation, but relatively small numbers of structures are involved in any one area. Other interface areas present the reverse image: isolated forest and wildland areas, usually in park lands, are interspersed between largely urban areas. Fires in these wildland areas rarely rage out of control.

The most significant wildland/urban interface problems are posed in areas where subdivisions and other developments abut wildland. This was the case in the Lake Wildwood subdivision in Nevada County, California, where the "49er" fire in 1988 caused extensive damage and earned itself the dubious distinction among California firefighters as "the fire of the 90s."[7]

---

[7] "Sunset Magazine Gives Tips for Living in Wildfire Country," *Wildfire News & Notes* (January/February 1990), p. 3.

In an interview with the *Washington Post*, fire prevention official Loren Poore of the CDF, noted that the 49er fire occurred in an area whose population has grown fourfold within the preceding ten to fifteen years. People subdivided large lots and "the problem got there before many people realized it," Poore said.

This trend toward rural living has enormous implications for firefighting. The strategies and techniques that are effective in fighting wildland fires are very different from those employed for the protection of lives and structures in residential areas. Because wildland firefighting and structural firefighting have been separate from each other, firefighters have not traditionally been cross-trained to handle the unique problems associated with each. Protecting structures in sparsely populated areas not only involves special logistical problems, it often stretches limited firefighting resources to the limit and may preclude those fighting the wildfire from taking the most effective fire suppression approach.

Measures to Minimize Fire Danger

With the growth of the wildland/urban interface, the once-clear distinction between areas of wildland and structural firefighting has begun to blur. Fire departments in these rapidly growing wildland/urban interface areas face tremendous challenges and are devising effective prevention strategies for meeting them.

In California, where fire problems along the wildland/urban interface are not new, a State law mandates certain fire prevention measures for residents in and around wildland areas. The law governs areas where the State has fire suppression responsibility, which includes all State- and privately-held forests and wildlands.

The California law requires homeowners to provide 30 feet of "defensible space" separating their homes and other structures from wooded areas. This area must be kept clear of brush and undergrowth. The law also allows fire officials the option of increasing that requirement to 100. Homeowners are also required to clear pine needles off roofs and to install screens over chimney openings.

Following the 1988 "49er fire," Ranger Bob Paulus of the California Department of Natural Resources and Fire Prevention reported that where the State requirements were observed, they appeared to effectively limit fire damages. In the Lake Wildwood subdivision, Paulus noted, most of the homes destroyed abutted the wildland area. While the fire did skip through the area, it did not seem to randomly spare some houses and gut others. Shake roofs were the biggest offenders, Paulus said. Houses with metal or composition roofs and Lake Wildwood subdivision homes that complied with the "defensible space" requirements in California State law were generally spared.[8]

Paulus said that while the provisions in the State law were generally sufficient, enforcement of those laws is often impossible. People continue to build without regard to fire safety, despite repeated warnings about the dangers involved. "There isn't one person out there who lost a house that hasn't received word from us about what to do," Paulus said. "They chose not to do anything about it."

---

[8] The same finding resulted in fires in Pebble Beach, California. See U.S. Fire Administration Fire Investigation Report, "Urban Wildlands Fire, Pebble Beach, California, May 31, 1987."

The Compliance Problem: Successful Approaches

The compliance problem is one that has been successfully addressed in some jurisdictions. The Los Angeles City Fire Department (LAFD) instituted a program that guarantees 100 percent compliance with brush clearance regulations by October of each year, when the wildfire danger is very high. LAFD created a system and authority to cite hazards, give a reasonable time for compliance with existing brush clearance laws, and then have the property cleared if the owner failed to do so, wrote Captain James O. Haworth, of the LAPD Brush Clearance Unit in the October 1989 issue of *Fire Journal*. Under Los Angeles' Brush Clearance Program, an ordinance gives the fire department the authority to hire private brush-clearance contractors to clear brush and vegetation hazards from private property if the owner fails to do so, and to bill the owner for doing so.

Since its implementation in 1981, 94 percent of homeowners have voluntarily complied with brush clearance laws after being notified of the regulations each April in the department's mass mailing to residents. Of the 9,000 parcels cited each year, Haworth reported that the department ends up contracting out only about 900, or about 10 percent. By October of each year, all properties are in compliance.

The small city of Orinda, California, took a similar approach and realized similar results. Using accepted advertising principles, the fire department carefully crafted an official letter about the need to clear weeds and brush. After extensive reader testing, the department sent the letter to its 7,000 home and lot owners. They have found the method very successful in combating non-compliance with weed abatement ordinances.

Florida is trying another approach to wildfire prevention. The Fire Control Bureau launched a Statewide fire reduction initiative in January 1989 in conjunction with the Forest Education Bureau. The Florida Fire Reduction Initiative coordinates a number of fire prevention programs, including traditional fire prevention, wildland/urban interface, and services of the Florida Forestry and Arson Alert Association. Included in the effort is a program designed to involve local citizens in neighborhood watch-type activities to protect their neighborhoods from careless wildland fires.

The LAPD, the Los Angeles County Fire Department, and the USFS Fire Lab have also been working on a project to test the efficacy of using prescribed fires as a wildland fire prevention technique in interface areas. They completed prescribed burns on an estimated 350 acres of heavy brush on hillsides directly below housing developments in the Stone Canyon. Haworth predicts that the Stone Canyon Research Project may prove the value of prescribed burning as an efficient, cost-effective fire management tool for fire protection agencies in interface areas.

Ongoing Initiatives

Following a disastrous wildfire year in 1985 in which 1,400 homes were lost, representatives of the USFS, USFA, and the NFPA, met to discuss the potential fire problems in the wildland/urban interface. A task force was formed which included representatives of those organizations as well as from insurance, architecture, wood products industry, academia, research organizations, and State and local government. The need for a national awareness effort was identified. Plans for a national meeting also came out of the task force's deliberations, and the first National Wildland/Urban Fire Protection Conference was held in September 1985.

The conference report, entitled "Wildfire Strikes Home!" concluded that the wildland/urban fire problem poses serious risks to life, property, and natural resources. That report cited low public awareness of the fire risk to homes in interface areas and recognized that the influx of new residents and homes was not likely to stop. The report also noted that while natural fire has always played a role in wildland areas, the influx of new residents was likely to increase the incidence and severity of the fires and to have major implications for firefighting efforts. Finally, the conference report concluded that while urban/wildland interface problems are likely to become national in scope, the involvement of regional and local players is necessary and essential to effectively implement national policy to meet local needs.

The effort that began in 1985 has continued to grow. Today, a variety of national and regional organizations are actively involved in issues related to the wildland/urban interface problem. The NWCG continues to address a number of issues related to interface problems. The National Association of State Foresters, whose members are directors of State forestry agencies in the 50 States and U.S. territories, has also played a leadership role in the field. In addition, a Wildland Fire Management Section has been established within the NFPA.

The National Association of State Foresters was the primary moving force behind legislation passed in 1990 that authorizes financial and technical assistance to volunteer fire departments in rural interface areas to purchase equipment and train firefighters in wildland firefighting methods. The "National Fire Forces Mobilization Act" authorizes up to 70 million dollars annually, subject to appropriations, to State forestry agencies and rural volunteer fire departments for activities necessary to enable their mobilization in the event of area, regional, and national fire emergencies.

Aftermath of the 1988 Wildfire Season: Lessons Learned for the Fire Service

The foregoing discussion presents an overview of the policies and issues brought into focus by the wildfires of 1988 and the wildland/urban interface problem. The discussion suggests that local firefighting organizations in areas near wildlands ought to be aware of what wildland fire management policies are, how those policies are made, and how they can be changed if change is necessary.

Since the "49er" fire, history has been made along the wildland/urban interface. In 1989, the Black Tiger Gulch fire just outside of Boulder was the worst ever in Colorado. The fire destroyed 44 structures and natural resources valued at 10 million dollars and burned for four days despite the efforts of an estimated 500 firefighters.

In June 1990, it was California. The price tag for damages was over 400 million dollars as a result of a wildfire that burned more than 600 structures and 4,900 acres and killed a woman in Santa Barbara County. It was the worst wildland/urban interface fire in California history, but only its scale was out of the ordinary.

As rural populations grow, the political pressure to concentrate efforts on protecting residents and structures is likely to increase. So too is the debate over Federal wildfire management policies. The mostly affluent new wildland dwellers are becoming more vocal in their demands for more and better fire protection. The 1988 fires in California and those in and around Yellowstone also brought strong and vocal criticism of "let burn" policies from nearby residents and business owners.

Fire service organizations in wildland/urban interface areas can influence the direction of USFS and NPS fire management plans and can implement wildland fire prevention measures to lessen the fire danger.

1. **Representatives of fire service organizations in areas near national forests, parks, and wilderness areas can affect wildfire management policies and practices by taking advantage of numerous opportunities for public participation in the development of fire management plans included in site-specific Forest Land Management Plans and National Park General Management Plans.**

Site-specific management plans for national forests and parks are updated periodically. Take advantage of opportunities to participate in the development of site-specific national forest or park management plans. Notices of hearings and opportunities for commenting on draft plans are published in *The Federal Register* and in local newspapers near affected parks and forests, and posted in U.S. post offices.

You can find out if a nearby forest or wilderness area already has an approved site-specific Forest Land Management Plan or if the park has an approved General Management Plan by contacting the forest ranger, Forest Service district office, or the local Park Service office.[9]

Even if there is no approved forest land or general management plan, you can obtain a copy of the fire management plans. Contact the district office for that national forest or the nearest regional office of the USFS or the NPS to find out how.

2. **Volunteer fire departments and fire protection districts in interface areas need to become more knowledgeable about wildland fire prevention and firefighting methods.**

Numerous opportunities exist for volunteers in rural departments and other interface areas to learn more about wildland fire prevention, protection, and suppression. (Some of these are described in Appendix A.)

3. **Fire departments can promote implementation of measures to help reduce fire dangers to homes and other structures in wildland/urban interface areas.**

A partial list of wildland fire prevention measures includes the following:[10]

The roof is the most vulnerable part of a building during a wildland fire. Shake roofs are the biggest offenders. Homes with metal or composition roofs are less vulnerable.

Automatic roof sprinkler systems will not substitute for safe roof material because of the unreliability of many rural water systems in a wildfire situation.

Vents, attic openings, foundation louvers, or other openings in vertical exterior walls and eave overhangs should not exceed 144 square inches and should be covered with 1.4 inch mesh metal screen that is noncombustible and corrosion-resistant.

Every chimney or vent attached to any solid or liquid fuel-burning device should be provided with an approved, securely attached spark arrestor that is visible from the ground.

---

[9] See Appendix B for a list of key Federal and regional fire management contacts for the USFS and NPS.

[10] The list included here is excerpted primarily from "Wildfire Strikes Home!," the report of the National Wildland/Urban Fire Protection Conference sponsored by the USFS, the NFPA, and the USFA, published by the NFPA in January 1987. Additional information was obtained through an interview with Bob Paulus of the CDF and from newspaper and magazine stories.

Exterior walls of buildings should be protected with materials of not less than one hour fire-resistant construction on the exterior side. Utility buildings of less than 100 square feet which are at least 50 feet from other buildings can be exempted.

All flammable vegetation should be cleared away from buildings for a minimum distance of 30 feet. Certain ornamental shrubbery, single specimens of trees, and other ground cover may be allowed in the fire break, provided they do not form a means of rapidly transmitting fire from the native growth to any structure.

Additional clearance, as far as 100 feet, may be necessary when extra hazardous conditions exist.

Overhanging trees within 10 feet of a chimney outlet should be cut back.

Any dead or dying wood should be removed from trees near a structure. Accumulated leaves, pine needles, and other dead vegetation should be removed from the roof.

4. **Fire departments in wildland/urban interface areas often need to take aggressive steps to implement and strictly enforce local weed abatement and brush clearance ordinances.**

Strong, carefully crafted letters to homeowners can be used effectively to spur compliance with brush clearance regulations.

Fire departments must have the authority to enforce compliance with these ordinances if the rules are to be effective in reducing the risk of wildfires in wildland/urban interface areas.

5. **Prescribed burning in wildland/urban interface areas can be a valuable, cost-effective fire management tool for fire protection agencies.**

The LAFD, the Los Angeles County Fire Department, and the USFS Fire Laboratory tested the efficacy of using carefully controlled fire service-ignited fires to prevent wildfires by burning heavy brush and other accumulated fuels in areas adjacent to residential areas.

APPENDIX A

Selected Available Resources

Wildland/Urban Interface Issues

Several wildland/urban interface resources for firefighters and fire service organizations have been developed by the Federal government and are available to fire service organizations. These include the following:

Wildland/Urban Interface Fire Protection: A National Problem with Local Solutions is a training kit developed by the USFA's National Fire Academy for volunteer fire departments and rural fire protection districts. The packet includes a textbook and workbook, a multiple choice exam, and a 46-minute video, and is designed for volunteers committed to wildland firefighting but who have little or no experience combating wildland residential fire hazards.

It is available for 30 dollars from the National AudioVisual Center, 8700 Edgeworth Drive, Capitol Heights, Maryland 20743-3701, 800-638-1300.

Protecting People and Homes From Wildfire In the Interior West contains the proceedings of a symposium and workshop held in Missoula, Montana, on a wide range of issues associated with protecting homes from wildfire. Single copies are available free from the USFS, Intermountain Research Station, Publications Office, 324 25th Street, Ogden, Utah 84401, 801-625-5437.

Wildland Fire Awareness in Your Community and *Wildland Fires and Your Home* are two new publications developed by the U.S. Department of Agriculture (USDA) Cooperative Extension at the University of Massachusetts. The publications provide information about the increased potential for life and property loss in the wildland/urban interface area and provide fire protection planning guidelines for the public, developers, builders, and land-use planners.

Contact the USDA Cooperative Extension, University of Massachusetts, Amherst, Massachusetts 01003, 413-545-2717.

Boise Interagency Fire Center Publications

The Boise Interagency Fire Center has a catalog of publications for fire service organizations. *The National Fire Equipment System Catalog* is a comprehensive guide to fire management equipment and publications available through the Boise Interagency Fire Center. While only Federal and State agencies may order equipment through the catalog, anyone with a legitimate need (including fire service organizations) can order publications. Also available from the Boise Interagency Fire Center are the *National Interagency Mobilization Guide* (NFES-2091) and regional mobilization guides for various areas of the country.

To order the publications catalog, write a letter on organizational letterhead requesting the *National Fire Equipment Catalog: Part 2 – Publications*. There is a nominal charge for the catalog for which you will be billed. No advance payments are accepted. Address your letters to Publications, Boise Interagency Fire Center, 3905 Vista Avenue, Boise, Idaho 83705. You may also call (208) 389-2542 for further information, but no phone orders are accepted.

APPENDIX B

Key Federal Fire Management Contacts

A. U.S. FIRE ADMINISTRATION

Tom Minnich
Associate Member, National Wildfire Coordinating Group
U.S. Fire Administration
16825 South Seton Avenue
Emmitsburg, Maryland 21727
301-447-1200

B. NATIONAL PARK SERVICE FIRE MANAGEMENT CONTACTS

National Park Service Headquarters
Elmer Hurd
Fire Director
U.S. Department of Interior
National Park Service
P.O. Box 37127, Room 3318
Washington, DC 20013-7127
202-208-6046

Regional Fire Management Officers

NPS Regional Office	*Fire Management Officer*
MID ATLANTIC REGION 143 South Third Street Philadelphia, Pennsylvania 19106	Doug Wallner 215-597-7140
MIDWEST REGION 1709 Jackson Street Omaha, Nebraska 68102	Ben Holmes 402-221-3475
NORTH ATLANTIC REGION 15 State Street Boston, Massachusetts 02109	Charisse Sydoriak
NATIONAL CAPITAL REGION 1100 Ohio Drive, SW Washington, DC 20242	Carl Douhan 202-619-7065
ROCKY MOUNTAIN REGION P.O. Box 25287 Denver, Colorado 80225	Tom Zimmerman 303-969-2449

B.   NATIONAL PARK SERVICE FIRE MANAGEMENT CONTACTS (continued)

Regional Fire Management Officers

NPS Regional Office	Fire Management Officer
PACIFIC NORTHWEST REGION 83 South King Street Suite 212 Seattle, Washington 98104	Mark Forbes 206-553-5670
SOUTHEAST REGION 75 Spring Street, SW Atlanta, Georgia 30303	Steve Smith 404-331-3527
SOUTHWEST REGION P.O. Box 728 Santa Fe, New Mexico 87501	Cliff Chetwin 505-988-6371
WESTERN REGION 600 Harrison Street Suite 600 San Francisco, California 94107	Chris Cameron 415-744-3921
ALASKA REGION 2525 Gambell Street Anchorage, Alaska 99503-2892	Steve Holder 907-257-2643

C.   U.S. FOREST SERVICE FIRE MANAGEMENT CONTACTS

U.S. Forest Service Headquarters

Lawrence A. Amicarella, Director
Fire and Aviation Management
U.S. Forest Service
1621 N. Kent Street
Arlington, Virginia 22209
202-453-9483

Regional Fire Management Contacts

National Forest Regions	Fire Management Contact
NORTHERN REGION Federal Building P.O. Box 7669 Missoula, Montana 59807	James F. Mann, Director Aviation & Fire Management 406-329-3402
John W. Mumma, Regional Forester 406-329-3316	

C. U.S. FOREST SERVICE FIRE MANAGEMENT CONTACTS (continued)

Regional Fire Management Contacts

National Forest Regions	*Fire Management Contact*
ROCKY MOUNTAIN REGION 11177 W. 8th Avenue P.O. Box 25127 Lakewood, Colorado 80225 Gary E. Cargill, Regional Forester 303-236-9427	Ray J. Evans, Director Air, Aviation & Fire Management 303-236-9641
SOUTHWESTERN REGION Federal Building 517 Gold Avenue, SW Albuquerque, New Mexico 87102 David F. Jolly, Regional Forester 505-842-3300	Jimmie Hickman, Director Aviation & Fire Management 505-842-3353
INTERMOUNTAIN REGION Federal Building 324 25th Street Ogden, Utah 84401 J.S. Tixier, Regional Forester 801-625-5605	Bill Price, Director Aviation & Fire Management 801-625-5507
PACIFIC SOUTHWEST REGION 630 Sansome Street San Francisco, California 94111 Paul F. Barker, Regional Forester 415-705-2870	Director Aviation & Fire Management 415-705-2788
PACIFIC NORTHWEST REGION 319 SW Pine Street P.O. Box 3623 Portland, Oregon 97208 John F. Butruille, Regional Forester 503-326-3625	Jim Bates, Director Aviation & Fire Management 503-326-2931
SOUTHERN REGION 1720 Peachtree Road, NW Atlanta, Georgia 30367 John E. Alcock, Regional Forester 404-347-4177	Dick A. Cox, Director Fire & Aviation 404-347-4243

C.   U.S. FOREST SERVICE FIRE MANAGEMENT CONTACTS (continued)

Regional Fire Management Contacts

National Forest Regions

Fire Management Contact

EASTERN REGION
310 W. Wisconsin Avenue
Room 500
Milwaukee, Wisconsin 53203

Floyd J. Marita, Regional Forester
414-291-3600

Richard Bacon, Director
Fire & Aviation
414-291-1898

ALASKA REGION
Federal Office Building
Box 21628
Juneau, Alaska 99802-1628

Michael A. Barton, Regional Forester
907-586-8863

Dennis Pendleton
Cooperative Fire
Management and NFS Fire Management
907-271-2575

www.ingramcontent.com/pod-product-compliance
Lightning Source LLC
Chambersburg PA
CBHW081239170526
45165CB00009B/3110